The Open University

Resource Book D

Units 14–16

This is the last of the four resource books which are part of the MU120 course materials. It contains further practice questions for *Units 14–16*, and revision questions for the whole course, together with solutions.

This publication forms part of an Open University course. Details of this and other Open University courses can be obtained from the Student Registration and Enquiry Service, The Open University, PO Box 197, Milton Keynes, MK7 6BJ, United Kingdom: tel. +44 (0)870 333 4340, e-mail general-enquiries@open.ac.uk

Alternatively, you may visit the Open University website at http://www.open.ac.uk where you can learn more about the wide range of courses and packs offered at all levels by The Open University.

To purchase a selection of Open University course materials, visit the webshop at www.ouw.co.uk, or contact Open University Worldwide, Michael Young Building, Walton Hall, Milton Keynes, MK7 6AA, United Kingdom, for a brochure: tel. +44 (0)1908 858785, fax +44 (0)1908 858787, e-mail ouwenq@open.ac.uk

The Open University, Walton Hall, Milton Keynes, MK7 6AA.

First published 1996. Third edition 2002. Reprinted 2003, 2005, 2006.

Copyright © 2002 The Open University

All rights reserved; no part of this publication may be reproduced, stored in a retrieval system, transmitted or utilised in any form or by any means, electronic, mechanical, photocopying, recording or otherwise, without written permission from the publisher or a licence from the Copyright Licensing Agency Ltd. Details of such licences (for reprographic reproduction) may be obtained from the Copyright Licensing Agency Ltd, 90 Tottenham Court Road, London W1T 4LP.

Open University course materials may also be made available in electronic formats for use by students of the University. All rights, including copyright and related rights and database rights, in electronic course materials and their contents are owned by or licensed to The Open University, or otherwise used by The Open University as permitted by applicable law.

In using electronic course materials and their contents you agree that your use will be solely for the purposes of following an Open University course of study or otherwise as licensed by The Open University or its assigns.

Except as permitted above you undertake not to copy, store in any medium (including electronic storage or use in a website), distribute, transmit or re-transmit, broadcast, modify or show in public such electronic materials in whole or in part without the prior written consent of The Open University or in accordance with the Copyright, Designs and Patents Act 1988.

Edited, designed and typeset by The Open University, using the Open University TeX System.

Printed and bound in the United Kingdom by The Charlesworth Group, Wakefield.

ISBN 0 7492 4468 2

Contents

Unit 14	Space and shape	4
Unit 15	Repeating patterns	7
Unit 16	Rainbow's end	10
Units 1–16	Revision	11
Solutions		21

Unit 14 Space and shape

Question 1

In each case, decide whether or not the shapes given are similar.

(a) The body of an adult and that of a toddler of the same sex.

(b) Any two chessboards.

(c) A right-angled triangle whose sides adjacent to the right angle are 4 cm and 5 cm long, and another right-angled triangle whose sides adjacent to the right angle are 12 cm and 15 cm long.

(d) The Earth and the Moon.

Question 2

(a) Figure 1 shows three horizontal lines AB, CD and EF. Which looks to be the shortest, and which the longest? Now measure all three lines. This is a well-known optical illusion. Try it on some other people if you can. Write a paragraph explaining why the illusion occurs.

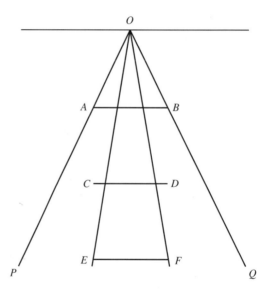

Figure 1

Question 3

A gym has an exercise machine that is a type of treadmill. The user can choose the gradient to walk or run up from a range of slopes, specified in the form '1 in n', where n can be any integer from 5 to 10. It is not clear, however, whether this means that the 'hill' rises 1 metre for every n metres along the slope, or for every n metres travelled horizontally. For both interpretations, make a table of the angle of inclination (in degrees) for the six values of n. Which interpretation gives the larger values of the slope?

Question 4

A right-angled triangle has sides adjacent to the right angle, which measure 4 and 5 units. What are the angles of the triangle?

Question 5

Look at Figure 2 below. A surveyor can find the height, h metres (CD), of a hill and the horizontal distance, x metres, from B to C by measuring the distance, d metres, from A to B and the angles α and β. Calculate the height of the hill when $d = 100$, $\alpha = 10°$ and $\beta = 15°$.

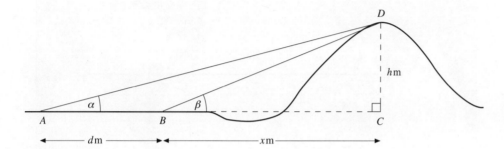

Figure 2

Question 6

While walking along the towpath at the edge of a straight canal, you decide you want to find out how wide the canal is. You have a compass, and you know your stride is about 1 metre. You notice a tree growing on the opposite bank a little way ahead. Using your compass, you measure the bearing of the tree from the direction of the canal edge: it is 50°. You take five strides back along the towpath and repeat the measurement: the new bearing is 30°. Draw a diagram and use trigonometry to estimate the width of the canal.

Your stride might be slightly more (or less) than 1 metre, so how accurate do you think your estimate is?

(Hint: this question is similar to Question 5.)

Question 7

The formula for the angle that subtends the great circle arc between two points on the globe can be written as

$$\theta = \cos^{-1}[\sin LT1 \sin LT2 + \cos LT1 \cos LT2 \cos(LG1 - LG2)].$$

An alternative that is sometimes used is

$$\theta = 90° - \sin^{-1}[\sin LT1 \sin LT2 + \cos LT1 \cos LT2 \cos(LG1 - LG2)].$$

Explain why these formulas are equivalent.

Question 8

In Figure 3, two horizontal lines, AB and CD, of equal length are drawn on a map of the Earth. Which line represents the larger distance on the Earth itself? Explain in a paragraph why equal distances on the map do not always correspond to equal distances on the surface of the Earth.

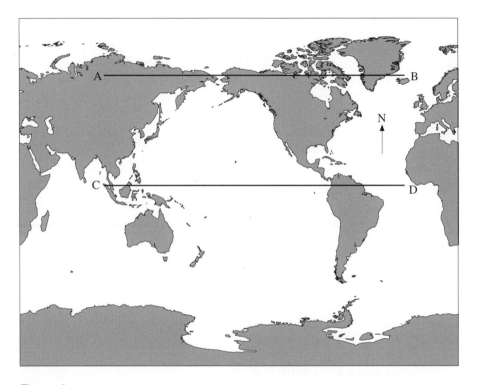

Figure 3

Question 9

The 'way points' listed in the table below were used by the Breitling Orbiter when making the first non-stop around-the-world balloon journey. Use the great circle calculator program, GRTCIRC, in Chapter 14.2 of the *Calculator Book* to calculate the distance (to the nearest km) between successive way points given in the table. Hence find the total distance travelled by the balloon.

Table 1

	Latitude	Longtitude
Start	46.5° N	7.1° E
Way point 1	19.0° N	42.5° E
Way point 2	25.0° N	106.0° E
Way point 3	9.0° N	163.8° W
Way point 4	16.1° N	85.4° W
Way point 5	23.1° N	18.9° W
Landing	26.2° N	28.4° E

How does this distance compare with the circumference of the Earth (40011 km)? Note: be careful to distinguish between east and west longitudes.

Unit 15 Repeating patterns

Question 1

The graph in Figure 4 shows the typical variation of the depth, d metres, of water in a particular harbour with time of day, t hours, as the depth changes with the tide.

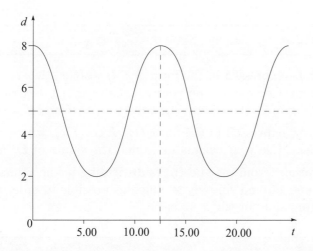

Figure 4

(a) (i) Write down the amplitude of the graph.

 (ii) Write down its period, and hence the frequency.

 (iii) Write down the phase shift relative to the standard sine curve.

 (iv) Hence find a function based on the sine function that will model the graph.

(b) On your calculator, plot the function that you have just found in (iv) above and use it to answer the following:

 (i) A boat enters the harbour at 3.30 a.m., after the high tide at midnight. The boat needs a water depth of 4 m to sail safely. What advice would you give the boat's captain about when to leave the harbour that afternoon if the boat is not to be forced to wait there through the evening low tide? Explain the reasons for your advice.

 (ii) State two modelling assumptions you have made.

Question 2

Two pure notes of the same amplitude and of frequencies 15 Hz and 17 Hz are sounded together to produce beats. Each note is modelled as a sine wave.

(a) If the function

$$Y_1 = \sin(at) + \sin(bt)$$

is to model the sound of the two notes played together, what values should the constants a and b take?

(b) Plot the function Y_1 on your calculator, choosing a suitable window setting to show the beats over a period of 1 second.

Question 3

In the UK, electricity is distributed as alternating current. The electrical voltage at a domestic mains socket can be modelled as a sine wave with a frequency of 50 Hz and an amplitude of about 340 V. What is the formula describing this sine wave?

(The figure of 240 V usually quoted for the mains voltage refers to the 'root-mean-square' value, which is $\dfrac{1}{\sqrt{2}}$ ($\simeq 0.707$) times the amplitude: this is a form of average value used in electrical engineering.)

Question 4

(a) $\cos^2 x$ can be expressed in the form $A + B\cos(Cx)$, where A, B and C are numbers. Enter $\cos^2 x$ on your calculator as $Y_1 = (\cos(X))^2$ and display the graph. Hence find the values of A, B and C.

(b) $\sin^2 x$ can be expressed in the form $D + E\cos(Fx)$, where D, E and F are numbers. Use your calculator to find the values of D, E and F.

(c) Add the results from (a) and (b) (substituting in the values of A, B, C, D, E and F) and simplify as much as possible in order to find a simple expression for $\sin^2 x + \cos^2 x$.

Question 5

Table 2 gives the times of sunrise and sunset in Liverpool at four-week intervals from early 1996 to early 1997. The times are in Greenwich Mean Time and are expressed in hours and minutes.

Use the data and the sine regression facility on your calculator to set up a mathematical model for predicting the length of daylight (from sunrise to sunset) at the beginning of each week in the period covered by Table 2.

Table 2 Sunrise and sunset in Liverpool from early 1996 to early 1997

Week	Date	Sunrise	Sunset	Week	Date	Sunrise	Sunset
0	6 Jan	08.26	16.09	28	20 Jul	04.09	20.26
4	3 Feb	07.54	16.57	32	17 Aug	04.55	19.35
8	2 Mar	06.57	17.52	36	14 Sept	05.44	18.12
12	30 Mar	05.49	18.44	40	12 Oct	06.33	17.22
16	27 Apr	04.45	19.35	44	9 Nov	07.26	16.24
20	25 May	03.57	20.21	48	7 Dec	08.13	15.54
24	22 Jun	03.43	20.44	52	4 Jan	08.26	16.09

Question 6

An alternative way of writing the trigonometric identity

$$\sin a + \sin b = 2\cos\left(\frac{a-b}{2}\right)\sin\left(\frac{a+b}{2}\right) \tag{1}$$

is as

$$\sin P + \sin Q = 2\cos X \sin Y, \tag{2}$$

where P and Q are expressions involving X and Y. What are those expressions? Write the identity (2) in terms of X and Y only.

Question 7

The Fourier series of a square wave has the form

$$A(\sin \omega t + \frac{1}{3}\sin 3\omega t + \frac{1}{5}\sin 5\omega t + \frac{1}{7}\sin 7\omega t + \cdots).$$

(a) If a square wave has a period of 1 second, what is the value of the fundamental frequency ω?

(b) The value of A can be chosen to fix the amplitude of the square wave. Use your calculator to estimate the value of A that will give a square wave with an amplitude of 1.

Unit 16 Rainbow's end

Question 1

Snell's law of refraction can be written as $\sin r = k \sin i$, where r and i are the angles of refraction and incidence, respectively.

(a) Use your calculator to plot a graph of r against i for the following data:

Angle of incidence i/degrees	10	20	30	40	50	60	70	80
Angle of refraction r/degrees	7.5	15	22	29	35	40.5	44.5	47.5

(b) Write Snell's law in the form $r =$ (an expression in k and i).

(c) Enter into your calculator the function you obtained for r in part (b) and store the value 0.5 for k. Plot the predicted graph of r for the incidence angles 10, 20, ... (in degrees), as in part (a). Now try various values of k between 0.5 and 1, and display the predicted graphs of r together with the measured data from the table. Which value of k gives the best fit between Snell's law and the experimental data? Compare your value with Descartes' value of 187/250.

Question 2

In *Paralipomena* (1604) and *Dioptrice* (1611), Kepler published his attempts at finding a trigonometric law of refraction. Some of his formulas, given below, include the difference, d, between the angle of incidence i and the angle of refraction r. Replace d by $i - r$ and rearrange each of the formulas into the more conventional form $r =$ (an expression in i and k):

(a) $d = ki$
(b) $d = k/\cos i$
(c) $\tan i = k \tan r$
(d) $\tan i = k \sin d$
(e) $1 - (\tan i / \tan d) = k \tan i$
(f) $1 - (\tan i / \tan d) = k \sin i$.

Question 3

In the primary rainbow model due to Descartes and Newton, the exit angle Y can be plotted against the impact parameter X for different colours. Each coloured band appears at an angle Y corresponding to the maximum point on its graph; this angle is sometimes called the Descartes' angle for the colour concerned. For any exit angle less than the Descartes' angle, there are *two* corresponding values of the impact parameter X.

(a) On your calculator, plot the graph of the exit angle Y against the impact parameter X in the case of red light, using $Y = 4\sin^{-1}(kX) - 2\sin^{-1} X$ with the value of k for red light ($k = 81/108$). (You may have already done this in Activity 16 of Unit 16.) Find the Descartes' angle and the associated impact parameter for red light by means of either the zoom or table facilities.

(b) Supernumerary red light (an additional faint arc just inside the primary bow) appears at an angle of about 40.1° (which is just below the angle for violet light in the primary bow). Use the graph from part (a) to find the two values of the impact parameter X that return red light at this angle.

Units 1–16   Revision

Question 1

Convert each of the following calculations into decimals (to two decimal places). Which **two** produce the same answer?

Options

A $\left(\dfrac{10}{7}\right)^2$ B $\dfrac{7}{4} + \dfrac{13}{25}$ C $\dfrac{30}{36.9}$ D $\sqrt{(3^2 + 4^2)} \times \dfrac{20}{7}$

E $\dfrac{300}{\sqrt{36}}$ F $\dfrac{4}{3} \times \dfrac{25}{7}$ G $\left(\dfrac{21}{300}\right)^{-1}$ H $^-2 + 1.11$

Question 2

One light year is the distance travelled by light in one year. This distance is approximately 9.5×10^{12} km. Which **four** options give the distance travelled by light in 1 second, correct to two significant figures? For the number of days in a year, use 365.

Options

A 3.0×10^8 km   B 3.0×10^8 m

C 300 million km   D 300 000 m

E 300 million m   F 3.0×10^5 km

G 3.0×10^{11} km   H 300 000 km

Question 3

The eight CMA scores of a student taking a 60-point Open University course are as follows:

| 80 | 78 | 83 | 94 | 75 | 66 | 86 | 63 |

(a) Select the option below that is closest to the mean of the scores in the table.

Options

A 82.4   B 77.6   C 78.1   D 78.7

E 76.7   F 79.9   G 81.3   H 79.2

(b) Select the option below that is closest to the median of the scores in the table.

Options

A 69   B 71   C 73   D 75

E 77   F 79   G 81   H 83

(c) Select the option below that is closest to the standard deviation of the scores in the table.

Options

A 3.1   B 4.7   C 6.3   D 9.5

E 10.2   F 12.7   G 15.1   H 19.8

Question 4

Select the **two** options that are true statements.

Options

A A price ratio of 1.5 is equivalent to a percentage price increase of 15%.

B A price ratio of 0.7 is equivalent to a percentage price decrease of 30%.

C A price ratio of 2.3 is equivalent to a percentage price increase of 23%.

D A percentage price increase of 5.5% is equivalent to a price ratio of 1.55.

E A percentage price decrease of 14% is equivalent to a price ratio of 0.86.

F A percentage price decrease of 20% is equivalent to a price ratio of 1.2.

Question 5

This question is based on the various prices for a computer given in Table 3.

Table 3 Computer price/£s

| 1440 | 1570 | 1470 | 1750 | 1430 | 1495 | 1555 | 1505 | 1395 | 1465 |

(a) Select the option below that corresponds to the range of the computer prices in Table 3.

Options

A 115   B 105   C 355   D 125

E 145   F 305   G 335   H None of these

(b) Select the option above that corresponds to the interquartile range of the computer prices in Table 3.

(c) Use your calculator to set up a boxplot for the data in Table 3, with a window for a range of X from 1300 to 1800. Choose the option opposite that most closely resembles the graph of the boxplot produced by the calculator.

Options

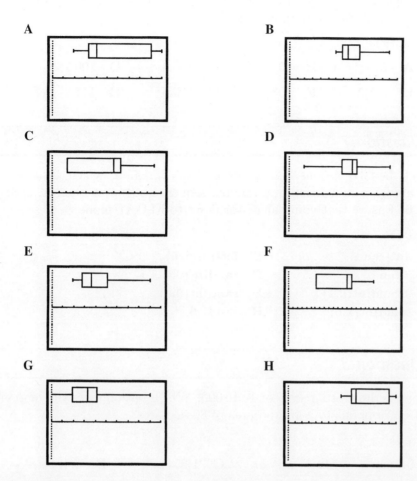

Question 6

In September 1997, the Retail Prices Index (RPI) stood at 159.3 (base date January 1987). The Average Earnings Index (AEI) for that month was 139.1 (base year 1990). In September 1996, the RPI was 153.8 and the AEI was 133.3.

(a) Select the option below that corresponds to the percentage increase in average earnings between September 1996 and September 1997 (rounded to one decimal place).

Options

A 0.9   **B** 1.0   **C** 2.4   **D** 3.4
E 4.3   **F** 4.4   **G** 5.8   **H** None of these

(b) In September 1996, the value of an index-linked pension was £88 per week, when adjusted for inflation. Which option below is closest to the value of the weekly pension, when adjusted for inflation, in September 1997?

Options

A £62   **B** £63   **C** £85   **D** £87
E £88   **F** £91   **G** £112   **H** £123

(c) Select the option below that is closest to the real earnings for September 1997 as a percentage of the real earnings one year earlier (rounded to one decimal place).

Options

A 73.1	**B** 99.3	**C** 100.4	**D** 100.7
E 103.4	**F** 103.5	**G** 116.8	**H** 117.2

Question 7

Select the **two** options that correspond to commands which, if used repeatedly, would generate a random sequence of the numbers 1, 2, 3, 4, 5 and 6. Ensure that your calculator is set to FLOAT mode.

Options

- **A** int(rand)
- **B** int(rand*6)
- **C** 6rand
- **D** randInt(0,5)+1
- **E** 6randInt(0,1)
- **F** randInt(0,6)
- **G** randInt(1,6)
- **H** rand1,6

Question 8

A map is drawn to a scale of 1 : 10 000. Which option gives the map area that corresponds to a ground area of 80 hectares?

Options

- **A** $0.8\,\text{cm}^2$
- **B** $1\,\text{cm}^2$
- **C** $1.25\,\text{cm}^2$
- **D** $6.4\,\text{cm}^2$
- **E** $8\,\text{cm}^2$
- **F** $10\,\text{cm}^2$
- **G** $64\,\text{cm}^2$
- **H** $80\,\text{cm}^2$

Question 9

Which **two** of the following statements are true?

Options

A In the third quadrant of a graph, the values of x are negative and the values of y are negative.

B The graph of the equation $y = 8x + 2$ passes through the point $(^-2, 14)$.

C The equation $8y = 2x + 8$ represents a straight-line graph with a slope of 2.

D The slope of a straight-line graph can be calculated by dividing the y-coordinate of any point on the line by the x-coordinate of that point.

E Velocity is the mathematical term for speed.

F A negative gradient means that a graph slopes down from right to left.

G Average speed is calculated by multiplying the distance travelled by the journey time.

H The graph of a directly proportional relationship has an intercept of zero.

Question 10

Two pure tones of the same amplitude but with frequencies 16 Hz and 18 Hz, respectively, are sounded together to produce beats. Each tone is modelled as a sine wave. Which **two** options model the resulting waveform?

Options

A $\sin 16t + \sin 18t$
B $\sin 4\pi t + \sin 36\pi t$
C $\sin 2t + \sin 34t$
D $\sin 32\pi t + \sin 36\pi t$
E $2\cos(16t)\sin(18t)$
F $2\cos(4\pi t)\sin(32\pi t)$
G $2\cos(2\pi t)\sin(34\pi t)$
H $2\cos(32\pi t)\sin(36\pi t)$

Question 11

Select the option that is equivalent to 225° in radians.

Options

A $\frac{2}{3}\pi$
B $\frac{3}{4}\pi$
C $\frac{3}{8}\pi$
D $\frac{5}{2}\pi$
E $\frac{5}{4}\pi$
F $\frac{5}{8}\pi$
G $\frac{7}{2}\pi$
H None of these

Question 12

The formula

$$y = a - 2bx^3$$

is rearranged to make x the subject. Select the option that gives the correct rearranged formula.

Options

A $x = \left(\dfrac{y-a}{2}\right) - \sqrt{b}$
B $x = (y-a)/\sqrt[3]{2b}$
C $x = \dfrac{b}{2}\sqrt[3]{y-a}$
D $x = \dfrac{\sqrt[3]{b(y-a)}}{2}$
E $x = \sqrt[3]{2(y-a)/2b}$
F $x = \sqrt[3]{\dfrac{a-y}{2b}}$
G $x = \sqrt[3]{\dfrac{y-a}{2b}}$
H None of these

Question 13

Use your calculator to solve the equation

$$5x^3 + 12x = 11.$$

Select the option that is closest to the solution.

Options

A 0.74
B 0.75
C 0.76
D 0.77
E 0.78
F 0.79
G 0.80
H 0.81

Question 14

Which **three** of the following statements are true?

Options

A The solution of the equation $10^x = 100$ is 10.

B An exponentially growing population with a doubling time of 20 years will increase by 50% in 10 years.

C The expression $\sqrt[3]{x^{12}}$ is another way of writing x^4.

D Each year, 1% of the atoms in a block of a radioactive substance decay. This means that the number of atoms in the block is decreasing exponentially.

E A monthly interest rate of 2% is equivalent to an APR of 24%.

F The surface area of a sphere is directly proportional to the square of its volume.

G The radius of a sphere is directly proportional to the cube of its volume.

H The radius of a sphere is directly proportional to the cube root of its volume.

Question 15

Table 4 shows how the duration of a £50 000 mortgage is related to the required monthly payments, provided the interest rate remains unchanged during the course of the mortgage.

Table 4 Mortgage repayments

Duration of mortgage in years (X)	25	21.5	18	15	12.5
Monthly payment in £s (Y)	361.12	386.12	411.12	461.12	511.12

Input the data into your calculator and use the regression facilities to find the equation for the curve of best fit, with the coefficients rounded to three significant figures.

Select the **four** options that are true statements.

Options

A Power regression gives the best fit.

B Linear regression gives the best fit.

C Exponential regression gives the best fit.

D The power regression equation is $Y = 1790 X^{-0.501}$.

E The power regression equation is $Y = 1791 X^{-0.5008}$.

F The exponential regression equation is $Y = 699 \times (0.973)^X$.

G Paying off the mortgage at £450 per month will take about 16 years.

H Paying off the mortgage at £500 per month will take 10 years.

Question 16

The function $y = 25 + 45\exp(-0.14t)$ can be used to model the way that the temperature, $y°C$, of water in a domestic hot-water tank varies with the time, t hours, since the immersion heater was switched off. Select the **two** options that are correct interpretations of this model.

Options

A The water is cooling towards a room temperature of 20°C.
B The water is cooling towards a room temperature of 25°C.
C The water is cooling towards a room temperature of 45°C.
D The water is cooling towards a room temperature of 70°C.
E The initial temperature of the water was 20°C.
F The initial temperature of the water was 25°C.
G The initial temperature of the water was 45°C.
H The initial temperature of the water was 70°C.

Question 17

The lengths of the sides of a triangle are 12, 8 and 16 units. Each of the options below gives the lengths of the sides of other triangles. Select the **two** options that correspond to triangles which are similar to the given triangle.

Options

A 5.5, 2.5, 3.75
B 4, 8, 6
C 2.5, 3.5, 4.5
D 5.4, 6.8, 3.4
E $\frac{1}{4}, \frac{3}{8}, \frac{1}{2}$
F 50, 75, 125
G 10, 5, 5

Question 18

Compass bearings of a cairn are taken from two points 1 km apart along a straight north–south path. After adjustment for magnetic variation, the bearings are 15° and 75°. Select the **two** options that are nearest to the distances (in metres) from the two points to the cairn.

Options

A 77
B 299
C 325
D 648
E 966
F 1092
G 1115
H 1301

Question 19

Bangkok is 13° 45′ N and 100° 35′ E, whilst New York is 40° 45′ N and 74°W. From these data, determine which option gives the great circle distance, to the nearest 1000 km, from Bangkok to New York. It is recommended that you use the calculator program GRTCIRC given in Chapter 14.2 of the *Calculator Book*.

Options

A 10 000   B 11 000   C 12 000   D 13 000

E 14 000   F 15 000   G 16 000   H 17 000

Question 20

The arc of the great circle joining Milton Keynes to Bangkok subtends an angle of about 100° at the centre of the Earth. Select the option that corresponds most closely to the distance between Milton Keynes and Bangkok when measured over the surface of a terrestrial globe of radius 1 m.

Options

A 0.5 m   B 0.9 m   C 1.1 m   D 1.4 m

E 1.7 m   F 1.9 m   G 2.2 m   H 2.7 m

Question 21

A sine wave has an amplitude of 20, a period of 0.5 s and a phase shift of 45°. Which option represents this sine wave?

Options

A $20\sin(2\pi t + \pi/4)$ B $20\sin(4\pi t + \pi/4)$ C $0.5\sin(20t + \pi/4)$

D $0.5\sin(0.5t + 45)$ E $0.5\sin(20\pi t + 45)$ F $20\sin(2t + \pi/4)$

G $20\sin(4t + \pi/4)$ H None of these

Question 22

Table 5 gives the times of sunrise in Glasgow at four-week intervals. The times are in Greenwich Mean Time and are expressed in hours and minutes.

Table 5 Sunrise in Glasgow

Week	Sunrise	Week	Sunrise
0	08.46	28	04.01
4	08.09	32	04.53
8	07.05	36	05.47
12	05.52	40	06.42
16	04.42	44	07.41
20	03.48	48	08.33
24	03.31	52	08.46

Select the option that provides the best fit to the sunrise data, where x represents the week number (round coefficients to two significant figures).

Options

A $6.2\sin(0.12x + 1.9) + 2.6$

B $6.2\sin(1.9x + 0.12) + 2.6$

C $2.6\sin(0.12x + 1.9) + 6.2$

D $2.6\sin(1.9x + 0.12) + 6.2$

E $1.9\sin(0.12x + 6.2) + 2.6$

F $0.12\sin(1.9x + 2.6) + 6.2$

Question 23

Select the **two** options that are identities and are therefore true for all values of x.

Options

A $\sin x = \sin(x + 5\pi)$ B $\cos(x + \pi/2) = {}^-\cos x$

C $\sin(x) = {}^-\sin({}^-x)$ D $\cos x = {}^-\sin(x + \pi/2)$

E $\sin(x + \pi) = \cos(x + 3\pi/2)$ F $\cos(x + 4\pi) = \cos(x + 2\pi)$

G $\sin({}^-x) = \sin x$ H $\cos({}^-x) = {}^-\cos x$

Question 24

Which **four** of the following statements about the mathematical model of the rainbow are true?

Options

A The angle of the primary bow is about 51°.

B The angle of the primary bow is about 42°.

C The secondary bow is formed by the reflection and refraction of sunlight.

D The primary bow is formed only by the refraction of sunlight.

E On entering a raindrop, red light is refracted more than violet light.

F The refractive index n for red light is less than that for violet light.

G The angle between the primary and secondary bows is about 9°.

H The angle between the primary and secondary bows is about 42°.

Question 25

Which **three** of the following are *incorrect* programming commands for the TI-83 calculator?

Options

A :DispGraph B :For(S,10,0,2)

C :If V<.175 D :Lbl 0

E :Disp L3,L2,A F :Fix 0.5

G :Plot1(Boxplot, L1) H :Input X,Y

Question 26

An MU120 student produced the following program to calculate cooking times for joints of meat.

Program : COOKING
:Input "K?", K
:Input "T?", T
:Disp (K × 2.2) × T

Which **four** of the following statements are true?

Options

A The program calculates the cooking time for a joint whose weight is given in kilograms when the recipe book gives the time per pound.

B The program calculates the cooking time for a joint whose weight is given in pounds when the recipe book gives the time per kilogram.

C The cooking time must be entered in minutes.

D The cooking time must be entered in hours.

E The brackets in the last line of the program are necessary.

F The brackets in the last line of the program are unnecessary.

G The program is based upon the assumption that weight in kilograms is 2.2 times weight in pounds.

H The program is based upon the assumption that cooking time is proportional to weight.

Solutions

Unit 14

1

(a) The relative sizes of parts of the body change as people age; in particular, a toddler's head is larger in relation to the rest of his or her body than an adult's. The bodies of an adult and a toddler are therefore not similar.

(b) Chessboards, being square, are all similar to one another. Moreover, since each chessboard is subdivided into the same number of smaller squares, chessboards are also similarly patterned.

(c) The hypotenuse (in cm) of the first triangle is $\sqrt{4^2 + 5^2} = \sqrt{41}$, while that of the second is $\sqrt{12^2 + 15^2} = \sqrt{369} = 3\sqrt{41}$. So the sides of the larger triangle are all three times the length of those of the smaller one. Hence the triangles are similar.

Alternatively, the ratio of the given pair of sides in one triangle is the same as the ratio of the corresponding sides in the other triangle (0.8), and the angle contained between those sides is the same (a right angle), so the triangles are similar.

(d) If the Earth and Moon were both perfect spheres, then they would be similar. However, the Earth is not perfectly spherical, but is flattened at its poles. For the two bodies to be similar, the Moon would have to have this same feature proportionally to the same degree, which seems highly unlikely. The Earth and the Moon are therefore not exactly similar. (Further investigation of reference books will confirm that this answer is correct.)

2

All three lines are actually the same length. However, most people interpret the figure as a perspective drawing of supposedly parallel lines: PO, EO, FO and QO, with a vanishing point O. Hence the line EF is interpreted as being nearest to the observer, and AB as being furthest away.

If this were a perspective drawing, a line that was actually as long as EF but was as far away as AB would be drawn shorter than EF. So AB (which is drawn the same length as EF) appears to be longer than EF. Also, because AB appears to represent a line of the same length as PQ, it is tempting to see it as about three times the length of EF.

3

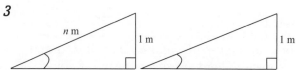

To find the angle corresponding to 'n metres along the slope', you must calculate $\sin^{-1}(1/n)$; for 'n metres horizontally', $\tan^{-1}(1/n)$ is required. You could use your calculator to find these values by entering the functions $Y_1 = \sin^{-1}(1/X)$ and $Y_2 = \tan^{-1}(1/X)$, and then using the table facility. The results (in degrees and radians, both correct to 3 s.f.) are given in the table below.

n	$\sin^{-1}(1/n)$	$\tan^{-1}(1/n)$
5	11.5° (0.201)	11.3° (0.197)
6	9.59° (0.167)	9.46° (0.165)
7	8.21° (0.143)	8.13° (0.142)
8	7.18° (0.125)	7.13° (0.124)
9	6.38° (0.111)	6.34° (0.111)
10	5.74° (0.100)	5.71° (0.100)

The table shows that $\sin^{-1}(1/n)$ always has the larger value. Hence the gradient is always greater when n is measured *along* the slope, although the difference decreases as n increases.

4

As this is a right-angled triangle, the angle opposite the side of length 4 is given by $\tan^{-1}(4/5) \simeq 38.66°$. The angle opposite the side of length 5 is therefore (approximately) $90° - 38.66° = 51.34°$.

5

Trigonometry can be used to find relationships between the variables in the two right-angled triangles ACD and BCD.

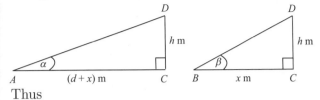

Thus

$$\tan \alpha = h/(d+x)$$

and

$$\tan \beta = \frac{h}{x}.$$

Algebra can be used to find h and x if d, α and β are known. There are several ways of doing the calculations. Here is one way.

As $d = 100$ and $\alpha = 10°$,

$$\tan 10° = \frac{h}{100+x},$$

so

$$(100+x)\tan 10° = h$$

or

$$h = 100\tan 10° + x\tan 10°. \qquad (1)$$

As $\beta = 15°$,

$$\tan 15° = \frac{h}{x},$$

so

$$h = x\tan 15°. \qquad (2)$$

Now (1) and (2) are both expressions for h and so must be equal. Therefore

$$100\tan 10° + x\tan 10° = x\tan 15°$$

$$100\tan 10° = x(\tan 15° - \tan 10°)$$

$$x = \frac{100\tan 10°}{(\tan 15° - \tan 10°)} = 192.45\ldots.$$

Substituting for x in (2) gives

$$h = 51.5668\ldots.$$

So the hill is about 52 m high (and B is about 190 m from C).

6 Call the point where the tree stands T. Let the point where you first measured the bearing be A, the point where you measured the second bearing be B, and let P be the point on the bank directly opposite the tree.

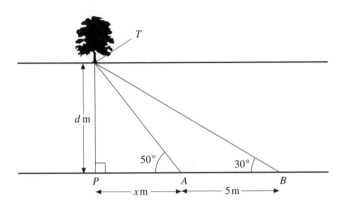

Let the width of the canal be d metres, and the distance AP be x metres. Then in triangle APT,

$$\tan 50° = \frac{d}{x},$$

so

$$x = d/\tan 50°, \qquad (3)$$

and in triangle BPT,

$$\tan 30° = \frac{d}{x+5},$$

so

$$x + 5 = \frac{d}{\tan 30°}. \qquad (4)$$

Substituting for x from (3) into (4) gives

$$\frac{d}{\tan 50°} + 5 = \frac{d}{\tan 30°},$$

so

$$5 = \frac{d}{\tan 30°} - \frac{d}{\tan 50°}$$

$$= d\left(\frac{1}{\tan 30°} - \frac{1}{\tan 50°}\right).$$

Hence

$$d = \frac{5}{(1/\tan 30° - 1/\tan 50°)}$$

$$\simeq 5.6.$$

Therefore the canal is about 5.6 m wide.

If your stride was such that the measurement of AB was 0.1 m more (or less), then AB would be 5.1 m (or 4.9 m) instead of 5 m. So

$$d = 5.1/(1/\tan 30° - 1/\tan 50°)$$

$$= 5.7.$$

Repeating the calculation with $AB = 4.9$ gives

$$d = 5.5.$$

Thus your answer is reasonably accurate for a rough estimate, and is fairly stable under slight changes in the assumptions. You may have considered a different 'error' in the measurement of your stride but reached the same conclusion.

Note: Although this question is similar to Question 5, a different method has been used in order to demonstrate an alternative approach.

7

For ease, call the expression in square brackets Y. So

$$Y = \sin LT1 \sin LT2 \\ + \cos LT1 \cos LT2 \cos(LG1 - LG2).$$

Then the first formula given becomes $\theta = \cos^{-1} Y$. Therefore, $\cos \theta = Y$.

One of the trigonometry identities given in Section 3 of *Unit 14* is

$$\cos \theta = \sin(90° - \theta).$$

This implies that

$$\sin(90° - \theta) = Y,$$

so

$$90° - \theta = \sin^{-1} Y.$$

Adding θ to both sides and subtracting $\sin^{-1} Y$ from both sides gives

$$\theta = 90 - \sin^{-1} Y.$$

Hence the two formulas for θ are equivalent.

8

The two lines are of equal length, but CD represents an east–west distance near the Equator, whereas AB represents an east–west distance near the Arctic. So CD represents a larger distance than AB, as explained below.

Because the Earth is a sphere, it cannot be represented accurately on a flat page. There has to be some distortion. Thus, the North and South Poles, which in reality are points on the Earth, are stretched to the full width of the map in Figure 3 (at the top and bottom, respectively). Hence east–west lines near the top and bottom of the map represent shorter distances than comparable lines near the Equator.

A similar effect holds for areas: Australia is not smaller than Greenland as it appears on this map, but over three times larger!

9

First point	Second point	Distance (to nearest km)
Start	Way point 1	4439 km
Way point 1	Way point 2	6521 km
Way point 2	Way point 3	9601 km
Way point 3	Way point 4	8497 km
Way point 4	Way point 5	6950 km
Way point 5	Landing	4764 km
Total		40772 km

The distance 40772 km is about 102% of the circumference of the Earth.

Unit 15

1

(a) (i) From the graph, the mean depth of water in the harbour is 5 m. The graph varies above and below this value, reaching a maximum of 8 and a minimum of 2. Therefore, the amplitude of the graph is $8 - 5 = 3$ (or $5 - 2 = 3$).

(ii) The period of the graph is 12.5 hours. Hence the frequency is $1/12.5 = 0.08\,\text{h}^{-1}$.

(iii) The standard sine curve would have started a quarter of a period to the left of the curve shown in the graph, and so the phase shift is $\frac{1}{4} \times 2\pi = \pi/2$ to the left (or $3\pi/2$ to the right).

(iv) The standard sine curve would have to be moved up by 5, as well as phase-shifted, in order to model the graph that represents the variation in harbour depth. Now, the standard sine curve can be written as $d = \sin(2\pi f t)$. So the function that models the graph is

$$d = 5 + 3\sin(2\pi \times 0.08t + \pi/2) \\ = 5 + 3\sin(0.16\pi t + \pi/2),$$

[or, $d = 5 + 3\sin(0.16\pi t - 3\pi/2)$].

(b) (i) Using TRACE or TABLE shows that the function falls to 4 m by about 16.30. So the boat needs to be well clear of the harbour by that time. Therefore, advise the pilot to aim to clear the harbour by at least 4 p.m. (16.00 hours).

(ii) The depth of water has been assumed to vary exactly as a sine curve, irrespective of the weather. It has also been assumed that the high tide is at precisely midnight.

2

(a) In the given function, the constants a and b correspond to the angular frequencies of the notes. So multiply the frequencies by 2π in order to get the angular frequencies. Therefore
$$a = 2\pi \times 15 = 30\pi \text{ and } b = 2\pi \times 17 = 34\pi.$$

(b) The sum of pure notes with frequencies 15 Hz and 17 Hz can be modelled by plotting $Y_1 = \sin(30\pi X) + \sin(34\pi X)$. The screen dump shown below was produced using the following window settings:

$X \min = 0, \ X \max = 1, \ X \text{ scl} = 1,$
$Y \min = ^-2, \ Y \max = 2, \ Y \text{ scl} = 1.$

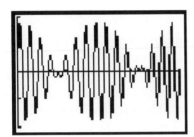

3

The general formula for a sine wave is $A\sin(\omega t + \phi)$, where A is the amplitude, ω is the angular frequency and ϕ is the phase shift. In this case, $A = 340$ V and $\omega = 2\pi f$, where the frequency $f = 50$ Hz, so $\omega = 100\pi$ (or about 314.2 radians per second). Take ϕ to be zero (since the time origin has not been specified). So the voltage V (in volts) at the mains socket is modelled by the formula
$$V = 340\sin(100\pi t).$$

4

(a) You should find that the graph of $\cos^2 x$ is itself a cosine curve, which varies above and below a mean value of 0.5, with an amplitude of 0.5. It completes two cycles for each complete cycle of $\cos x$. So $A = 0.5$, $B = 0.5$ and $C = 2$, giving the following trigonometric identity:
$$\cos^2 x = 0.5 + 0.5\cos 2x.$$

You can check this by plotting the functions $Y_1 = \cos^2 X$ and $Y_2 = 0.5 + 0.5\cos 2X$ on your calculator.

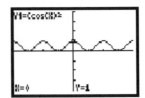 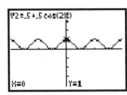

(b) You should find that the graph of $\sin^2 x$ also varies above and below a mean value of 0.5, with an amplitude of 0.5, and that it completes two cycles for each complete cycle of $\sin x$. This time, however, the curve is the negative of a cosine curve. So $D = 0.5$, $E = ^-0.5$ and $F = 2$, giving the following identity:
$$\sin^2 x = 0.5 - 0.5\cos 2x.$$

(c) Adding the two curves together gives
$$\sin^2 x + \cos^2 x$$
$$= 0.5 - 0.5\cos 2x + 0.5 + 0.5\cos 2x$$
$$= 1.$$

This is an important general result, as it is true for *all* values of x. See *Unit 15*, Activity 17.

5

The data given in Table 2 cover a period of 52 weeks.

The length of daylight is found by subtracting the time of sunrise from the time of sunset. This can be done by entering the data into lists on the calculator (remembering to convert from hours and minutes to a decimal) and using the list arithmetic operations. Then use sine regression (Sin Reg) on your calculator to obtain the model for the length of daylight y (hours) in terms of the week x:
$$y = 4.60\sin(0.12x - 1.25) + 12.20$$
(coefficients to 2 d.p.).

6

Comparing the two identities gives $P = a$, $Q = b$, $X = (a-b)/2$ and $Y = (a+b)/2$.

Substituting for a and b in the expressions for X and Y yields
$$X = \frac{P-Q}{2} \text{ and } Y = \frac{P+Q}{2}.$$

Rearranging this first expression to get P in terms of X and Q gives
$$2X = P - Q,$$

so
$$P = 2X + Q.$$
Rearranging the second expression
$$Y = \frac{P+Q}{2}$$
gives
$$2Y = P + Q.$$
Substituting for P yields
$$2Y = (2X + Q) + Q$$
$$= 2X + 2Q$$
$$Y = X + Q,$$
so
$$Q = Y - X.$$
Since
$$P = 2X + Q,$$
it follows that
$$P = 2X + Y - X$$
$$= X + Y.$$
Therefore, identity (2) can be written as
$$\sin(Y + X) + \sin(Y - X) = 2\cos(X)\sin(Y).$$

7

(a) Angular frequency ω is related to the period T by the formula $\omega = 2\pi/T$. Since $T = 1\,\text{s}$ for the square wave, the fundamental frequency $\omega = 2\pi/1 \simeq 6.263$ radians per second (to 3 d.p.).

(b) If you set $A = 1$ and use your calculator program FOURIER to plot the square wave from the Fourier series, you should find that the amplitude (that is, the height of the flat top of the waveform) approaches a value of about 0.785. To obtain an amplitude of 1, therefore, the Fourier series must be scaled by the factor $A = 1/0.785 = 1.274$.

[It turns out that the value of A can be calculated from the theory of Fourier series. It is directly proportional to the amplitude, and, for an amplitude X, the value of A is $4X/\pi$. Thus, for a square wave of amplitude 1, the value of A is $4/\pi$, or 1.273 (to 3 d.p.).]

Unit 16

1

(a) The answer is provided by your calculator screen and should look like the screen dump below.

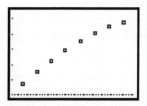

(b) Snell's law can be written in the form:
$$r = \sin^{-1}(k \sin i).$$
Note: this is *not* the same as $\sin^{-1} k \cdot \sin i$.

(c) You should find that a value of $k \simeq 0.75$ gives the best fit. Descartes' value is 0.748.

2

Substituting for d, as appropriate, and rearranging gives the following.

(a) $i - r = ki$,
so
$$r = i - ki \text{ or } i(1 - k).$$

(b) $i - r = k/\cos i$,
so
$$r = i - k/\cos i.$$

(c) $\tan r = (\tan i)/k$,
so
$$r = \tan^{-1}[(\tan i)/k].$$

(d) $\tan i = k \sin(i - r)$
$$\sin(i - r) = (\tan i)/k$$
$$i - r = \sin^{-1}[(\tan i)/k],$$
so
$$r = i - \sin^{-1}[(\tan i)/k].$$

(e) $1 - [\tan i/\tan(i - r)] = k \tan i$
$$1 - k \tan i = \tan i/\tan(i - r)$$
$$\tan(i - r) = \frac{\tan i}{1 - k \tan i}$$
$$i - r = \tan^{-1}\left(\frac{\tan i}{1 - k \tan i}\right),$$
so
$$r = i - \tan^{-1}\left(\frac{\tan i}{1 - k \tan i}\right).$$

(f) As in (e), but replace $k \tan i$ by $k \sin i$, so
$$r = i - \tan^{-1}\left(\frac{\tan i}{1 - k \sin i}\right).$$

3

(a)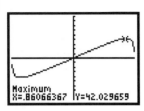

For red light, the Descartes' angle is 42.03° (0.73356 radian) and the associated impact parameter X is 0.861.

(b) Red light is returned at 40.1° by impact parameters $X = 0.761$ and 0.929.

Units 1–16

1

A 2.04 (to 2 d.p.)   B 2.27 (to 2 d.p.)
C 0.81 (to 2 d.p.)   D 14.29 (to 2 d.p.)
E 50   F 4.76 (to 2 d.p.)
G 14.29 (to 2 d.p.) H $^-0.89$

Options D and G produce the same answer.

Reference: *Calculator Book*, Chapter 1.2, 1.4 and 1.7.

2

Number of seconds in a year
$$= 365 \times 24 \times 60 \times 60.$$

Therefore, distance travelled by light in 1 second
$$= 9.5 \times 10^{12} \div (365 \times 24 \times 60 \times 60)\,\text{km}$$
$$= 301243.0238\,\text{km}$$
$$= 300\,000\,\text{km (to 2 s.f.) (option H)}$$
$$= 3 \times 10^5\,\text{km (option F)}$$
$$= 3 \times 10^5 \times 10^3\,\text{m} = 3 \times 10^8\,\text{m (option B)}$$
$$= 300 \times 10^6\,\text{m} = 300\,\text{million m (option E)}.$$

Options B, E, F and H are correct.

Reference: *Calculator Book*, Chapter 1.6.

3 Enter the scores in a list in your calculator and use 1-Var Stats (on STAT Calc).

(a) Mean, $\bar{x} = 78.125$.

Option C is correct.

(b) Median $= 79$.

Option F is correct.

(c) Standard deviation, $\sigma_x = 9.53$.

Option D is correct.

Reference: *Calculator Book*, Chapters 2.1 and 4.3.

4

A Not true: a price ratio of 1.5 is equivalent to a price increase of 50%.
B True
C Not true: a price ratio of 2.3 is equivalent to a price increase of 130%.
D Not true: a price increase of 5.5% is equivalent to a price ratio of 1.055.
E True
F Not true: a price *decrease* of 20% is equivalent to a price ratio of 0.8.

Option B and E are true.

Reference: *Unit 2*, Section 4.

5

(a) Inputting the prices into a list in your calculator and ordering them (or plotting a boxplot) gives the highest price (in £s) as 1750 and the lowest as 1395. So the range of computer prices is $1750 - 1395 = 355$.

Option C is correct.

(b) The quartiles are 1440 and 1555. So the interquartile range is $1555 - 1440 = 115$.

Option A is correct.

(c) The boxplot (with the specified window) looks like that in option E.

Reference: *Calculator Book*, Chapter 3.

SOLUTIONS

6

(a) The ratio of the relevant average earnings is
$$\frac{\text{AEI (Sept 1997)}}{\text{AEI (Sept 1996)}} = \frac{139.1}{133.3} = 1.04351.$$

As a percentage, this is 104.4% (to 1 d.p.), which corresponds to an increase of 4.4% in average earnings over the period considered.

Option F is correct.

(b) The price ratio that is relevant for the inflation adjustment here is
$$\frac{\text{RPI (Sept 1997)}}{\text{RPI (Sept 1996)}} = \frac{159.3}{153.8} \; (= 1.03576\ldots).$$

The index-linked pension will be increased by this proportion and so will be
$$£88 \times \frac{159.3}{153.8} = £91.1\ldots$$
$$= £91 \text{ (to the nearest £)}.$$

Option F is correct.

(c) Real earnings take into account the RPI. Thus
$$\frac{\text{real earnings (Sept 1997)}}{\text{real earnings (Sept 1996)}}$$
$$= \frac{\text{AEI (Sept 1997)}}{\text{AEI (Sept 1996)}} \times \frac{\text{RPI (Sept 1996)}}{\text{RPI (Sept 1997)}}$$
$$= \frac{139.1}{133.3} \times \frac{153.8}{159.3} = 1.00748\ldots.$$

As a percentage, this is 100.7% (to 1 d.p.). So, the real earnings in September 1997 will be approximately 100.7% of those a year earlier.

Option D is correct.

Reference: *Unit 3*, Section 7.

7

A Incorrect: always produces 0.
B Incorrect: produces a random sequence of the integers 0 to 5.
C Incorrect: does not produce integers.
D Correct
E Incorrect: produces a random sequence of just 0 and 6.
F Incorrect: produces a random sequence of 0 to 6.
G Correct
H Incorrect: gives a syntax error.

Options D and G are correct.

Reference: *Calculator Book*, Chapter 4.2.

8

One hectare is 100 m by 100 m.

On the map, 1 cm represents 10 000 cm or 100 m. So 1 cm^2 represents 100 m by 100 m, or 1 hectare. Therefore, 80 hectares are represented by 80 cm^2.

Alternatively, 80 hectares = $80 \times 100 \times 100$ m^2.

On the map, this is represented by
$$\frac{80 \times 100 \times 100 \text{ m}^2}{(10000)^2} = \frac{80}{10000} \text{ m}^2$$
$$= \frac{80}{10000} \times 100 \times 100 \text{ cm}^2$$
$$= 80 \text{ cm}^2.$$

Option H is correct.

Reference: *Unit 6*, Section 4.4.

9

A True
B Not true: if $x = {}^-2$, then $y = 8x + 2 = {}^-16 + 2 = {}^-14$ (not 14).
C Not true: $8y = 2x + 8$ does represent a straight-line graph, but rewriting the equation as $y = \tfrac{1}{4}x + 1$ shows that its slope is $\tfrac{1}{4}$.
D Not true: this holds only if the graph passes through the origin.
E Not true: velocity has a direction, whereas speed does not.
F Not true: the graph slopes down from left to right.
G Not true: *divide* the distance travelled by the journey time, rather than multiply.
H True: the graph passes through (0,0) and hence the intercept is 0.

Options A and H are true.

Reference: *Unit 7*, Section 1.3, and *Unit 10*, Section 1.

10

A pure tone of frequency 16 Hz can be modelled by $\sin(16 \times 2\pi t) = \sin 32\pi t$, and similarly a pure tone of frequency 18 Hz by $\sin 36\pi t$. So the resulting waveform is given by
$$\sin 32\pi t + \sin 36\pi t.$$

Option D is correct.

Plotting the functions on the same graph shows that this is the same function as

$$2\cos(2\pi t)\sin(34\pi t).$$

Option G is correct.

Reference: *Calculator Book*, Chapter 9.2.

11

As $180° = \pi$ radians, $1° = \dfrac{\pi}{180}$ radian.

So $225° = 225 \times \dfrac{\pi}{180} = \dfrac{25\pi}{20} = \dfrac{5\pi}{4}$ radians.

Option E is correct.

Reference: *Unit 9*, Section 1, *Unit 14* Section 1 and *Unit 15*, Section 1.

12

From

$$y = a - 2bx^3,$$
$$2bx^3 = a - y.$$

Hence

$$x^3 = \dfrac{a-y}{2b},$$

or

$$x = \sqrt[3]{\dfrac{a-y}{2b}}.$$

Option F is correct.

Reference: *Unit 8*, Section 4.

13

Enter $Y = 5x^3 + 12x$ on your calculator and find where $Y = 11$ is on the graph or table. Alternatively, enter $Y = 5x^3 + 12x - 11$ and find where $Y = 0$ is on the graph or table.

In both cases you should find that the solution occurs where $x = 0.7446$ (to 4 s.f.).

Option A is correct.

Reference: *Calculator Book*, Chapter 8.5.

14

A Not true: the solution is $x = 2$ since $10^2 = 100$.

B Not true: if a population increased by 50% every 10 years, it would be $(1.5)^2 = 2.25$ times larger after 20 years and so would have *more* than doubled.

C True: $x^{12} = x^4 \times x^4 \times x^4$, so $\sqrt[3]{x^{12}} = x^4$.

D True

E Not true: a monthly interest rate of 2% corresponds to 1.02 per month, and $(1.02)^{12} = 1.268$ per year (12 months) or an APR of 26.8%.

F Not true: the surface area of a sphere is directly proportional to the square of the radius *not* the square of the volume.

G Not true: the volume is directly proportional to the cube of the radius *not* vice versa.

H True

Options C, D and H are true.

Reference: *Unit 12*, Sections 4 and 5, and *Unit 13*, Section 1.

15

Put the duration (X) of the mortgage in list L1 and the monthly payments (Y) in list L2, then power regression gives $Y = aX^b$,

$$\text{where} \quad a = 1790.535278$$
$$\text{and} \quad b = {}^-0.5007843558,$$

with regression coefficient $r = {}^-0.9946748577$. Rounding a and b to 3 s.f. gives

$$Y = 1790X^{-0.501}.$$

Option D is true, and option E is not rounded as required.

Linear regression gives $Y = aX + b$,

$$\text{where} \quad a = 641.586\,3992$$
$$\text{and} \quad b = {}^-11.71013039,$$

with regression coefficient $r = {}^-0.9710135917$.

Exponential regression gives $Y = ab^X$,

$$\text{where} \quad a = 699.1256213$$
$$\text{and} \quad b = 0.9730370621,$$

with regression coefficient $r = {}^-0.9809837926$. Rounding a and b to 3 s.f. gives

$$Y = 699 \times (0.973)^X.$$

So option F is true.

Comparing regression coefficients suggests that power regression gives the best fit. So option A is true, and options B and C are not.

Putting the power regression function into $Y1$ and using either the Trace or Table facility gives $X = 15.764$ when $Y = 450$, and $X = 12.773$ when $Y = 500$. So option G is true, and option H is not, as it ignores the interest payments. (Also, from the table, payments of £500 per month will correspond to a mortgage of duration between 12.5 and 15 years.)

Therefore options A, D, F and G are true.

Reference: *Calculator Book*, Chapters 10.1, 12.4, 13.1 and 13.2.

16

When $t = 0$, $y = 25 + 45 = 70$. So option H is correct.

As t increases, $\exp(^-0.14t)$ tends to zero, and so y tends to 25. Hence option B is correct.

(Note: exp is entered using the e^x button on the calculator.)

Reference: *Calculator Book*, Chapter 12.3, *Unit 12*, Section 6, and *Unit 13*, Sections 4–6.

17

For two triangles to be similar, the ratios of the lengths of the sides in one triangle must be the same as the ratios of the lengths of the corresponding sides in the other triangle. In the given triangle the length of the longest side is twice the length of the shortest, and the intermediate-sized side is 1.5 times the shortest.

The sides in option B have these ratios.

The sides in option E also have these ratios.

None of the other options have these ratios.

(Note that it is easiest to compare ratios if the lengths in the triangles are all arranged in the same order—for example, from longest to shortest.)

Options B and E are correct.

Reference: *Unit 14*, Section 1.

18

In the diagram below, A and B represent the points from which bearings are taken, and C represents the cairn.

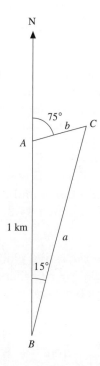

In triangle ABC, angle $A = 180° - 175° = 105°$ and angle $B = 15°$, so
angle $C = 180° - 105° - 15° = 60°$. Application of the sine formula to triangle ABC with the distances in kilometres ($AB = 1$) gives
$$\frac{a}{\sin 105°} = \frac{b}{\sin 15°} = \frac{1}{\sin 60°}.$$
So
$$a = \frac{\sin 105°}{\sin 60°} = 1.1153\ldots$$
and
$$b = \frac{\sin 15°}{\sin 60°} = 0.2988\ldots.$$
These are distances in kilometres. Multiply by 1000 to get the distances in metres. Hence the distances from the two points to the cairn are 1115 m and 299 m (to the nearest whole number).

Options B and G are correct.

Reference: *Unit 14*, Section 4.

19

Using the notation in the *Calculator Book*, Chapter 14.2,

$$LT1 = 13°45' \, \text{N} = 13 + 45/60,$$
$$LG1 = 100°35' \, \text{E} = 100 + 35/60,$$
$$LT2 = 40°45' \, \text{W} = 40 + 45/60,$$
$$LG2 = 74° \, \text{W} = {}^-74°.$$

Employ the calculator program GRTCIRC (make sure your calculator is in degree mode) and find the great circle distance from Bangkok to New York, which is $13955.74452 \simeq 14\,000$ km (to the nearest 1000 km).

Option E is correct.

Reference: *Unit 14*, Section 5 and Appendix, and *Calculator Book*, Chapter 14.2.

20

The distance, l, on a terrestrial globe is given by the arc length formula $l = r\theta$, where θ is the subtended angle in radians and r is the radius of the globe.

Now $100° = \dfrac{100}{180}\pi$ radians and $r = 1$ m, so

$$\text{distance on globe} = 1 \times \frac{100}{180}\pi$$
$$\simeq 1.7 \, \text{m (to 2 s.f.)}.$$

Option E is correct.

Reference: *Unit 14*, Section 1.

21

A sine wave $A\sin(wt + \phi)$ has amplitude A, period $2\pi/w$ and phase shift ϕ.

In this case, $A = 20$, $2\pi/w = 0.5$ (so $\omega = 4\pi$), and $\phi = \dfrac{45\pi}{180} = \dfrac{\pi}{4}$.

Therefore the given sine wave is represented by

$$20\sin(4\pi t + \pi/4).$$

Option B is correct.

Reference: *Unit 15*, Section 2, and *Calculator Book*, Chapter 15.1.

22

Application of the procedure in the *Calculator Book*, Section 15.1, gives the time of sunrise, y, as $y = a\sin(bx + c) + d$, where $a = 2.56\ldots$, $b = 0.1158\ldots$, $c = 1.895\ldots$, $d = 6.22\ldots$.

So, with coefficients rounded to 2 s.f.,

$$y = 2.6\sin(0.12x + 1.9) + 6.2.$$

Option C is correct.

Reference: *Calculator Book*, Section 15.1.

23

Try substituting some values of x into the options to find out which options are true. For each option, you might also plot the graphs corresponding to the functions on either side of the equals sign to see if they coincide.

A Not true (try $x = \pi/2$, for example).
B Not true (try $x = 0$, for example).
C True: the sine graph for negative values is $({}^-1)$ times that for positive values.
D Not true (try $x = 0$, for example).
E Not true (try $x = \pi/2$, for example).
F True: the cosine graph repeats every 2π.
G Not true (try $x = \pi/2$, for example).
H Not true (try $x = 0$, for example).

Options C and F are true.

Reference: *Unit 15*, Section 2.

24

A Not true: the angle of the primary bow is about 42°.
B True
C True
D Not true: reflection is also involved.
E Not true: violet light is refracted more than red light.
F True: for red light $k = 81/108$, so $n = 108/81$. For violet light $k = 81/109$, so $n = 109/81$.
G True: the angle of the secondary bow is about 51°, which is 9° more than that of the primary bow.
H Not true

Options B, C, F, and G are true.

Reference: *Unit 16*, Section 2.

25

Use Appendix A of the calculator manual if you are unsure.

- **A** Correct
- **B** Incorrect: you cannot move up from 10 to 0 in steps of +2 (it is the wrong direction).
- **C** Correct
- **D** Correct
- **E** Correct
- **F** Incorrect: you cannot have 0.5 decimal places.
- **G** Correct
- **H** Incorrect: you can only input one variable at a time (unlike **Prompt**).

Options B, F and H are incorrect.

Reference: *Calculator Book*, Chapter 16.

26

The program requires the input of **K**, the weight of the joint, which it multiplies by 2.2. (and then by the time **T**). So the program converts kilograms to pounds before multiplying by the time per pound, **T**, which can be entered in any units.

Options A and G are true.

The brackets in the last line are not needed as
$(K \times 2.2) \times T = K \times 2.2 \times T$.

So option F is true.

The program is based on the assumption that the cooking time is a constant multiplied by the weight.

So option H is also true.

Reference: *Calculator Book*, Chapter 16.